ALGEBRA 2 / TRIGONOMETRY REGENTS REVIEW

Published by
TOPICAL REVIEW BOOK COMPANY
P. O. Box 328
Onsted, MI 49265-0328

ALGEBRA 2 / TRIGONOMETRY REGENTS REVIEW

Published by
TOPICAL REVIEW BOOK COMPANY
P. O. Box 328
Onsted, MI 49265-0328

EXAM	PAGE
Note To Student	i
Reference Sheet	ii
June 2010	1
August 2010	9
January 2011	17
June 2011	26
January 2012	34
June 2012	42
January 2013	50

NOTE TO STUDENT

A straightedge (ruler) and a graphing calculator must be available for the exclusive use of each student taking the examination. The memory of any calculator with programming capability must be cleared, reset, or disabled. Students may not use calculators that are capable of symbol manipulation or that can communicate with other calculators through infrared sensors, nor may students use operating manuals, instruction or formula cards, or other information concerning the operation of calculators during the examination.

Algebra 2/Trigonometry Reference Sheet

Area of a Triangle
$K = \dfrac{1}{2} ab \sin C$

Functions of the Sum of Two Angles
$\sin(A + B) = \sin A \cos B + \cos A \sin B$
$\cos(A + B) = \cos A \cos B - \sin A \sin B$
$\tan(A + B) = \dfrac{\tan A + \tan B}{1 - \tan A \tan B}$

Functions of the Difference of Two Angles
$\sin(A - B) = \sin A \cos B - \cos A \sin B$
$\cos(A - B) = \cos A \cos B + \sin A \sin B$
$\tan(A - B) = \dfrac{\tan A - \tan B}{1 + \tan A \tan B}$

Law of Sines
$\dfrac{a}{\sin A} = \dfrac{b}{\sin B} = \dfrac{c}{\sin C}$

Sum of a Finite Arithmetic Series
$S_n = \dfrac{n(a_1 + a_n)}{2}$

Binomial Theorem
$(a + b)^n = {}_nC_0 a^n b^0 + {}_nC_1 a^{n-1} b^1 + {}_nC_2 a^{n-2} b^2 + \ldots + {}_nC_n a^0 b^n$
$(a + b)^n = \displaystyle\sum_{r=0}^{n} {}_nC_r a^{n-r} b^r$

Law of Cosines
$a^2 = b^2 + c^2 - 2bc \cos A$

Functions of the Double Angle
$\sin 2A = 2 \sin A \cos A$
$\cos 2A = \cos^2 A - \sin^2 A$
$\cos 2A = 2 \cos^2 A - 1$
$\cos 2A = 1 - 2 \sin^2 A$
$\tan 2A = \dfrac{2 \tan A}{1 - \tan^2 A}$

Functions of the Half Angle
$\sin \dfrac{1}{2}A = \pm \sqrt{\dfrac{1 - \cos A}{2}}$
$\cos \dfrac{1}{2}A = \pm \sqrt{\dfrac{1 + \cos A}{2}}$
$\tan \dfrac{1}{2}A = \pm \sqrt{\dfrac{1 - \cos A}{1 + \cos A}}$

Sum of a Finite Geometric Series
$S_n = \dfrac{a_1(1 - r^n)}{1 - r}$

Normal Curve
Standard Deviation

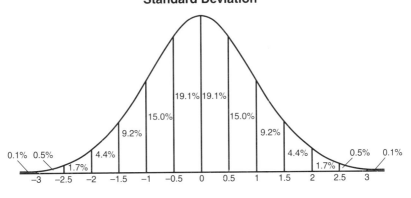

ii

ALGEBRA 2 / TRIGONOMETRY
June 2010
Part I

Answer all 27 questions in this part. Each correct answer will receive 2 credits. No partial credit will be allowed. For each question, record your answer, using a No. 2 pencil, in the space provided. [54]

1. What is the common difference of the arithmetic sequence 5, 8, 11, 14?
(1) $\frac{8}{5}$ (2) -3 (3) 3   (4) 9   1 _____

2. What is the number of degrees in an angle whose radian measure is $\frac{11\pi}{12}$?
(1) 150   (2) 165   (3) 330   (4) 518   2 _____

3. If $a = 3$ and $b = -2$, what is the value of the expression $\frac{a^{-2}}{b^{-3}}$?
(1) $-\frac{9}{8}$ (2) -1 (3) $-\frac{8}{9}$ (4) $\frac{8}{9}$ 3 _____

4. Four points on the graph of the function f(x) are shown below.
$$\{(0, 1), (1, 2), (2, 4), (3, 8)\}$$
Which equation represents f(x)?
(1) $f(x) = 2^x$ (2) $f(x) = 2x$ (3) $f(x) = x + 1$ (4) $f(x) = \log_2 x$ 4 _____

5. The graph of $y = f(x)$ is shown to the right.

Which set lists all the real solutions of $f(x) = 0$?

(1) $\{-3, 2\}$
(2) $\{-2, 3\}$
(3) $\{-3, 0, 2\}$
(4) $\{-2, 0, 3\}$

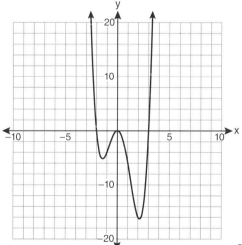

5 _____

6. In simplest form, $\sqrt{-300}$ is equivalent to
(1) $3i\sqrt{10}$ (2) $5i\sqrt{12}$ (3) $10i\sqrt{3}$ (4) $12i\sqrt{5}$ 6 _____

ALGEBRA 2 / TRIGONOMETRY
June 2010

7. Twenty different cameras will be assigned to several boxes. Three cameras will be randomly selected and assigned to box A. Which expression can be used to calculate the number of ways that three cameras can be assigned to box A?

(1) $20!$ (2) $\dfrac{20!}{3!}$ (3) $_{20}C_3$ (4) $_{20}P_3$

8. Factored completely, the expression $12x^4 + 10x^3 - 12x^2$ is equivalent to
(1) $x^2(4x + 6)(3x - 2)$
(2) $2(2x^2 + 3x)(3x^2 - 2x)$
(3) $2x^2(2x - 3)(3x + 2)$
(4) $2x^2(2x + 3)(3x - 2)$

9. The solutions of the equation $y^2 - 3y = 9$ are
(1) $\dfrac{3 \pm 3i\sqrt{3}}{2}$ (2) $\dfrac{3 \pm 3i\sqrt{5}}{2}$ (3) $\dfrac{-3 \pm 3\sqrt{5}}{2}$ (4) $\dfrac{3 \pm 3\sqrt{5}}{2}$

10. The expression $2 \log x - (3 \log y + \log z)$ is equivalent to
(1) $\log \dfrac{x^2}{y^3 z}$ (2) $\log \dfrac{x^2 z}{y^3}$ (3) $\log \dfrac{2x}{3yz}$ (4) $\log \dfrac{2xz}{3y}$

11. The expression $(x^2 - 1)^{-\frac{2}{3}}$ is equivalent to
(1) $\sqrt[3]{(x^2 - 1)^2}$ (2) $\dfrac{1}{\sqrt[3]{(x^2 - 1)^2}}$ (3) $\sqrt{(x^2 - 1)^3}$ (4) $\dfrac{1}{\sqrt{(x^2 - 1)^3}}$

12. Which expression is equivalent to $\dfrac{\sqrt{3} + 5}{\sqrt{3} - 5}$?
(1) $-\dfrac{14 + 5\sqrt{3}}{11}$ (2) $-\dfrac{17 + 5\sqrt{3}}{11}$ (3) $\dfrac{14 + 5\sqrt{3}}{14}$ (4) $\dfrac{17 + 5\sqrt{3}}{14}$

13. Which relation is *not* a function?
(1) $(x - 2)^2 + y^2 = 4$ (2) $x^2 + 4x + y = 4$ (3) $x + y = 4$ (4) $xy = 4$

14. If $\angle A$ is acute and $\tan A = \dfrac{2}{3}$, then
(1) $\cot A = \dfrac{2}{3}$
(2) $\cot A = \dfrac{1}{3}$
(3) $\cot(90° - A) = \dfrac{2}{3}$
(4) $\cot(90° - A) = \dfrac{1}{3}$

15. The solution set of $4^{x^2 + 4x} = 2^{-6}$ is
(1) $\{1, 3\}$ (2) $\{-1, 3\}$ (3) $\{-1, -3\}$ (4) $\{1, -3\}$

16. The equation $x^2 + y^2 - 2x + 6y + 3 = 0$ is equivalent to
(1) $(x - 1)^2 + (y + 3)^2 = -3$ (3) $(x + 1)^2 + (y + 3)^2 = 7$
(2) $(x - 1)^2 + (y + 3)^2 = 7$ (4) $(x + 1)^2 + (y + 3)^2 = 10$ 16 _____

17. Which graph best represents the inequality $y + 6 \geq x^2 - x$?

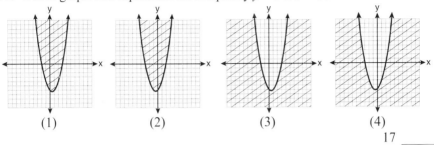

(1)   (2)   (3)   (4)
                                                                                17 _____

18. The solution set of the equation $\sqrt{x + 3} = 3 - x$ is
(1) {1}   (2) {0}   (3) {1, 6}   (4) {2, 3}   18 _____

19. The product of i^7 and i^5 is equivalent to
(1) 1   (2) –1   (3) i (4) $-i$ 19 _____

20. Which equation is represented by the accompanying graph?
(1) $y = \cot x$
(2) $y = \csc x$
(3) $y = \sec x$
(4) $y = \tan x$

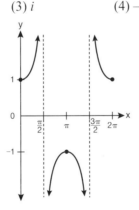

20 _____

21. Which value of r represents data with a strong negative linear correlation between two variables?
(1) –1.07   (2) –0.89   (3) –0.14   (4) 0.92   21 _____

22. The function $f(x) = \tan x$ is defined in such a way that $f^{-1}(x)$ is a function. What can be the domain of $f(x)$?
(1) $\{x \mid 0 \leq x \leq \pi \}$ (3) $\{x \mid -\frac{\pi}{2} < x < \frac{\pi}{2}\}$
(2) $\{x \mid 0 \leq x \leq 2\pi \}$ (4) $\{x \mid \neg \frac{\pi}{2} < x < \frac{3\pi}{2}\}$ 22 _____

23. In the accompanying diagram of right triangle KTW, KW = 6, KT = 5, and m∠KTW = 90. What is the measure of ∠K, to the *nearest minute*?

(1) 33°33' (3) 33°55'
(2) 33°34' (4) 33°56'

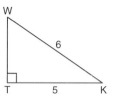

23 _____

24. The expression $\cos^2 \theta - \cos 2\theta$ is equivalent to
(1) $\sin^2 \theta$ (2) $-\sin^2 \theta$ (3) $\cos^2 \theta + 1$ (4) $-\cos^2 \theta - 1$ 24 _____

25. Mrs. Hill asked her students to express the sum $1 + 3 + 5 + 7 + 9 + \ldots + 39$ using sigma notation. Four different student answers were given. Which student answer is correct?

(1) $\sum_{k=1}^{20}(2k-1)$ (2) $\sum_{k=2}^{40}(k-1)$ (3) $\sum_{k=-1}^{37}(k+2)$ (4) $\sum_{k=1}^{39}(2k+1)$ 25 _____

26. What is the formula for the nth term of the sequence 54, 18, 6, ...?

(1) $a_n = 6\left(\frac{1}{3}\right)^n$ (2) $a_n = 6\left(\frac{1}{3}\right)^{n-1}$ (3) $a_n = 54\left(\frac{1}{3}\right)^n$ (4) $a_n = 54\left(\frac{1}{3}\right)^{n-1}$ 26 _____

27. What is the period of the function $y = \frac{1}{2}\sin\left(\frac{x}{3} - \pi\right)$?

(1) $\frac{1}{2}$ (2) $\frac{1}{3}$ (3) $\frac{2}{3}\pi$ (4) 6π 27 _____

Part II

Answer all 8 questions in this part. Each correct answer will receive 2 credits. Clearly indicate the necessary steps, including appropriate formula substitutions, diagrams, graphs, charts, etc. For all questions in this part, a correct numerical answer with no work shown will receive only 1 credit. All answers should be written in pen, except for graphs and drawings, which should be done in pencil. [16]

28. Use the discriminant to determine all values of k that would result in the equation $x^2 - kx + 4 = 0$ having equal roots.

29. The scores of one class on the Unit 2 mathematics test are shown in the accompanying table.

Find the population standard deviation of these scores, to the *nearest tenth*.

Unit 2 Mathematics Test

Test Score	Frequency
96	1
92	2
84	5
80	3
76	6
72	3
68	2

30. Find the sum and product of the roots of the equation $5x^2 + 11x - 3 = 0$.

31. The graph of the equation $y = \left(\frac{1}{2}\right)^x$ has an asymptote. On the grid below, sketch the graph of $y = \left(\frac{1}{2}\right)^x$ and write the equation of this asymptote.

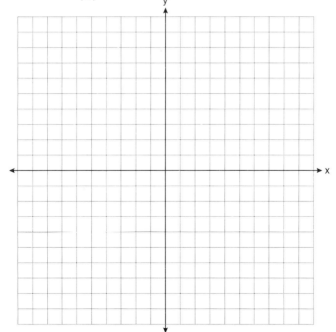

32. Express $5\sqrt{3x^3} - 2\sqrt{27x^3}$ in simplest radical form.

33. On the unit circle shown in the accompanying diagram, sketch an angle, in standard position, whose degree measure is 240 and find the exact value of sin 240°.

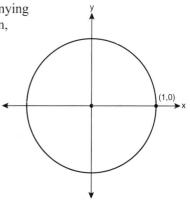

34. Two sides of a parallelogram are 24 feet and 30 feet. The measure of the angle between these sides is 57°. Find the area of the parallelogram, to the *nearest square foot*.

35. Express in simplest form: $\dfrac{\dfrac{1}{2} - \dfrac{4}{d}}{\dfrac{1}{d} + \dfrac{3}{2d}}$

ALGEBRA 2 / TRIGONOMETRY
June 2010
Part III

Answer all 3 questions in this part. Each correct answer will receive 4 credits. Clearly indicate the necessary steps, including appropriate formula substitutions, diagrams, graphs, charts, etc. For all questions in this part, a correct numerical answer with no work shown will receive only 1 credit. All answers should be written in pen, except for graphs and drawings, which should be done in pencil. [12]

36. The members of a men's club have a choice of wearing black or red vests to their club meetings. A study done over a period of many years determined that the percentage of black vests worn is 60%. If there are 10 men at a club meeting on a given night, what is the probability, to the *nearest thousandth*, that *at least* 8 of the vests worn will be black?

37. Find all values of θ in the interval 0° ≤ θ < 360° that satisfy the equation sin 2θ = sin θ.

38. The letters of any word can be rearranged. Carol believes that the number of different 9-letter arrangements of the word "TENNESSEE" is greater than the number of different 7-letter arrangements of the word "VERMONT." Is she correct? Justify your answer.

ALGEBRA 2 / TRIGONOMETRY
June 2010
Part IV

Answer the question in this part. A correct answer will receive 6 credits. Clearly indicate the necessary steps, including appropriate formula substitutions, diagrams, graphs, charts, etc. A correct numerical answer with no work shown will receive only 1 credit. The answer should be written in pen. [6]

39. In a triangle, two sides that measure 6 cm and 10 cm form an angle that measures 80°. Find, to the *nearest degree*, the measure of the smallest angle in the triangle.

ALGEBRA 2 / TRIGONOMETRY

August 2010
Part I

Answer all 27 questions in this part. Each correct answer will receive 2 credits. No partial credit will be allowed. For each question, write in the space provided the numeral preceding the word or expression that best completes the statement or answers the question. [54]

1. The product of $(3 + \sqrt{5})$ and $(3 - \sqrt{5})$ is
(1) $4 - 6\sqrt{5}$ (2) $14 - 6\sqrt{5}$ (3) 14   (4) 4   1 _____

2. What is the radian measure of an angle whose measure is $-420°$?
(1) $-\dfrac{7\pi}{3}$ (2) $-\dfrac{7\pi}{6}$ (3) $\dfrac{7\pi}{6}$ (4) $\dfrac{7\pi}{3}$ 2 _____

3. What are the domain and the range of the function shown in the accompanying graph?

(1) $\{x \mid x > -4\}; \{y \mid y > 2\}$
(2) $\{x \mid x \geq -4\}; \{y \mid y \geq 2\}$
(3) $\{x \mid x > 2\}; \{y \mid y > -4\}$
(4) $\{x \mid x \geq 2\}; \{y \mid y \geq -4\}$

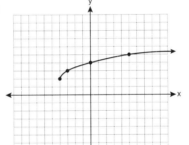

3 _____

4. The expression $2i^2 + 3i^3$ is equivalent to
(1) $-2 - 3i$ (2) $2 - 3i$ (3) $-2 + 3i$ (4) $2 + 3i$ 4 _____

5. In which graph is θ coterminal with an angle of $-70°$?

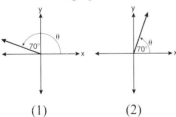 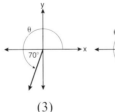 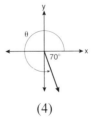

(1)   (2)   (3)   (4)   5 _____

6. In $\triangle ABC$, $m\angle A = 74$, $a = 59.2$, and $c = 60.3$. What are the two possible values for $m\angle C$, to the *nearest tenth*?
(1) 73.7 and 106.3   (3) 78.3 and 101.7
(2) 73.7 and 163.7   (4) 78.3 and 168.3   6 _____

7. What is the principal value of $\cos^{-1}\left(-\dfrac{\sqrt{3}}{2}\right)$?
(1) $-30°$ (2) $60°$ (3) $150°$ (4) $240°$ 7 _____

8. What is the value of x in the equation $9^{3x+1} = 27^{x+2}$?
(1) 1 (2) $\frac{1}{3}$ (3) $\frac{1}{2}$ (4) $\frac{4}{3}$

9. The roots of the equation $2x^2 + 7x - 3 = 0$ are
(1) $-\frac{1}{2}$ and -3 (2) $\frac{1}{2}$ and 3 (3) $\frac{-7 \pm \sqrt{73}}{4}$ (4) $\frac{7 \pm \sqrt{73}}{4}$

10. Which ratio represents csc A in the accompanying diagram?
(1) $\frac{25}{24}$ (2) $\frac{25}{7}$ (3) $\frac{24}{7}$ (4) $\frac{7}{24}$

11. When simplified, the expression $\left(\frac{w^{-5}}{w^{-9}}\right)^{\frac{1}{2}}$ is equivalent to
(1) w^{-7} (2) w^2 (3) w^7 (4) w^{14}

12. The principal would like to assemble a committee of 8 students from the 15-member student council. How many different committees can be chosen?
(1) 120 (2) 6,435 (3) 32,432,400 (4) 259,459,200

13. An amateur bowler calculated his bowling average for the season. If the data are normally distributed, about how many of his 50 games were within one standard deviation of the mean?
(1) 14 (2) 17 (3) 34 (4) 48

14. What is a formula for the nth term of sequence B shown below?
$$B = 10, 12, 14, 16, \ldots$$
(1) $b_n = 8 + 2n$ (3) $b_n = 10(2)^n$
(2) $b_n = 10 + 2n$ (4) $b_n = 10(2)^{n-1}$

15. Which values of x are in the solution set of the following system of equations?
$$y = 3x - 6$$
$$y = x^2 - x - 6$$
(1) 0, −4 (2) 0, 4 (3) 6, −2 (4) −6, 2

16. The roots of the equation $9x^2 + 3x - 4 = 0$ are
(1) imaginary
(2) real, rational, and equal
(3) real, rational, and unequal
(4) real, irrational, and unequal

17. In $\triangle ABC$, $a = 3$, $b = 5$, and $c = 7$. What is m$\angle C$?
(1) 22 (2) 38 (3) 60 (4) 120

18. When $x^{-1} - 1$ is divided by $x - 1$, the quotient is
(1) -1 (2) $-\dfrac{1}{x}$ (3) $\dfrac{1}{x^2}$ (4) $\dfrac{1}{(x-1)^2}$ 18 _____

19. The fraction $\dfrac{3}{\sqrt{3a^2b}}$ is equivalent to

(1) $\dfrac{1}{a\sqrt{b}}$ (2) $\dfrac{\sqrt{b}}{ab}$ (3) $\dfrac{\sqrt{3b}}{ab}$ (4) $\dfrac{\sqrt{3}}{a}$ 19 _____

20. Which graph represents a one-to-one function?

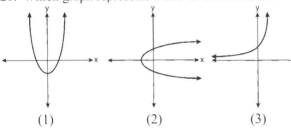

(1)   (2)   (3)   (4)   20 _____

21. The sides of a parallelogram measure 10 cm and 18 cm. One angle of the parallelogram measures 46 degrees. What is the area of the parallelogram, to the *nearest square centimeter*?
(1) 65   (2) 125   (3) 129   (4) 162   21 _____

22. The minimum point on the graph of the equation $y = f(x)$ is $(-1, -3)$. What is the minimum point on the graph of the equation $y = f(x) + 5$?
(1) $(-1, 2)$ (2) $(-1, -8)$ (3) $(4, -3)$ (4) $(-6, -3)$ 22 _____

23. The graph of
$y = x^3 - 4x^2 + x + 6$
is shown to the right.

What is the product of the roots of the equation
$x^3 - 4x^2 + x + 6 = 0$?
(1) -36
(2) -6
(3) 6
(4) 4

23 _____

24. What is the conjugate of $-2 + 3i$?
(1) $-3 + 2i$ (2) $-2 - 3i$ (3) $2 - 3i$ (4) $3 + 2i$ 24 _____

25. What is the common ratio of the geometric sequence whose first term is 27 and fourth term is 64?
(1) $\frac{3}{4}$ (2) $\frac{64}{81}$ (3) $\frac{4}{3}$ (4) $\frac{37}{3}$ 25 _____

26. Which graph represents one complete cycle of the equation $y = \sin 3\pi x$?

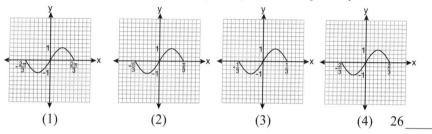

(1)   (2)   (3)   (4)   26 _____

27. Which two functions are inverse functions of each other?
(1) $f(x) = \sin x$ and $g(x) = \cos x$
(2) $f(x) = 3 + 8x$ and $g(x) = 3 - 8x$
(3) $f(x) = e^x$ and $g(x) = \ln x$
(4) $f(x) = 2x - 4$ and $g(x) = -\frac{1}{2}x + 4$ 27 _____

Part II

Answer all **8** questions in this part. Each correct answer will receive **2** credits. Clearly indicate the necessary steps, including appropriate formula substitutions, diagrams, graphs, charts, etc. For all questions in this part, a correct numerical answer with no work shown will receive only 1 credit. All answers should be written in pen, except for graphs and drawings, which should be done in pencil. [16]

28. Factor completely: $10ax^2 - 23ax - 5a$

29. Express the sum $7 + 14 + 21 + 28 + \ldots + 105$ using sigma notation.

30. Howard collected fish eggs from a pond behind his house so he could determine whether sunlight had an effect on how many of the eggs hatched. After he collected the eggs, he divided them into two tanks. He put both tanks outside near the pond, and he covered one of the tanks with a box to block out all sunlight. State whether Howard's investigation was an example of a controlled experiment, an observation, or a survey. Justify your response.

31. The accompanying table shows the number of new stores in a coffee shop chain that opened during the years 1986 through 1994.

Using $x = 1$ to represent the year 1986 and y to represent the number of new stores, write the exponential regression equation for these data. Round all values to the *nearest thousandth*.

Year	Number of New Stores
1986	14
1987	27
1988	48
1989	80
1990	110
1991	153
1992	261
1993	403
1994	681

32. Solve the equation $2 \tan C - 3 = 3 \tan C - 4$ algebraically for all values of C in the interval $0° \leq C < 360°$.

33. A circle shown in the diagram below has a center of $(-5, 3)$ and passes through point $(-1, 7)$.

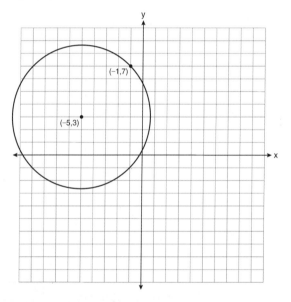

Write an equation that represents the circle.

34. Express $\left(\frac{2}{3}x - 1\right)^2$ as a trinomial.

35. Find the total number of different twelve-letter arrangements that can be formed using the letters in the word *PENNSYLVANIA*.

Part III

Answer all 3 questions in this part. Each correct answer will receive 4 credits. Clearly indicate the necessary steps, including appropriate formula substitutions, diagrams, graphs, charts, etc. For all questions in this part, a correct numerical answer with no work shown will receive only 1 credit. All answers should be written in pen, except for graphs and drawings, which should be done in pencil. [12]

36. Solve algebraically for x: $\dfrac{1}{x+3} - \dfrac{2}{3-x} = \dfrac{4}{x^2-9}$

37. If $\tan A = \frac{2}{3}$ and $\sin B = \frac{5}{\sqrt{41}}$ and angles A and B are in Quadrant I, find the value of $\tan (A + B)$.

38. A study shows that 35% of the fish caught in a local lake had high levels of mercury. Suppose that 10 fish were caught from this lake. Find, to the *nearest tenth of a percent*, the probability that *at least* 8 of the 10 fish caught did *not* contain high levels of mercury.

Part IV
Answer the question in this part. A correct answer will receive 6 credits. Clearly indicate the necessary steps, including appropriate formula substitutions, diagrams, graphs, charts, etc. A correct numerical answer with no work shown will receive only 1 credit. [6]

39. Solve algebraically for x: $\log_{x+3} \frac{x^3 + x - 2}{x} = 2$

ALGEBRA 2 / TRIGONOMETRY

17

January 2011
Part I

Answer all 27 questions in this part. Each correct answer will receive 2 credits. No partial credit will be allowed. For each question, write in the space provided the numeral preceding the word or expression that best completes the statement or answers the question. [54]

1. Which graph does *not* represent a function?

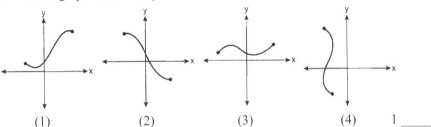

(1)   (2)   (3)   (4)   1 _____

2. The roots of the equation $x^2 - 10x + 25 = 0$ are
(1) imaginary
(2) real and irrational
(3) real, rational, and equal
(4) real, rational, and unequal   2 _____

3. Which values of x are solutions of the equation $x^3 + x^2 - 2x = 0$?
(1) 0, 1, 2   (2) 0, 1, –2   (3) 0, –1, 2   (4) 0, –1, –2   3 _____

4. In the accompanying diagram of a unit circle, the ordered pair $\left(-\dfrac{\sqrt{2}}{2}, -\dfrac{\sqrt{2}}{2}\right)$ represents the point where the terminal side of θ intersects the unit circle.

What is m$\angle\theta$?
(1) 45   (3) 225
(2) 135   (4) 240

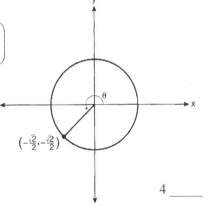

4 _____

5. What is the fifteenth term of the sequence 5, –10, 20, –40, 80, ... ?
(1) –163,840   (2) –81,920   (3) 81,920   (4) 327,680   5 _____

6. What is the solution set of the equation $|4a + 6| - 4a = -10$?
(1) ∅   (2) {0}   (3) {$\dfrac{1}{2}$}   (4) {0, $\dfrac{1}{2}$}   6 _____

7. If $\sin A = \dfrac{2}{3}$ where $0° < A < 90°$, what is the value of $\sin 2A$?
(1) $\dfrac{2\sqrt{5}}{3}$ (2) $\dfrac{2\sqrt{5}}{9}$ (3) $\dfrac{4\sqrt{5}}{9}$ (4) $-\dfrac{4\sqrt{5}}{9}$ 7 _____

8. A dartboard is shown in the accompanying diagram. The two lines intersect at the center of the circle, and the central angle in sector 2 measures $\dfrac{2\pi}{3}$. If darts thrown at this board are equally likely to land anywhere on the board, what is the probability that a dart that hits the board will land in either sector 1 or sector 3?

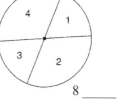

(1) $\dfrac{1}{6}$ (2) $\dfrac{1}{3}$ (3) $\dfrac{1}{2}$ (4) $\dfrac{2}{3}$ 8 _____

9. If $f(x) = x^2 - 5$ and $g(x) = 6x$, then $g(f(x))$ is equal to
(1) $6x^3 - 30x$ (2) $6x^2 - 30$ (3) $36x^2 - 5$ (4) $x^2 + 6x - 5$ 9 _____

10. Which arithmetic sequence has a common difference of 4?
(1) $\{0, 4n, 8n, 12n, ...\}$
(2) $\{n, 4n, 16n, 64n, ...\}$
(3) $\{n + 1, n + 5, n + 9, n + 13, ...\}$
(4) $\{n + 4, n + 16, n + 64, n + 256, ...\}$ 10 _____

11. The conjugate of $7 - 5i$ is
(1) $-7 - 5i$ (2) $-7 + 5i$ (3) $7 - 5i$ (4) $7 + 5i$ 11 _____

12. If $\sin^{-1}\left(\dfrac{5}{8}\right) = A$, then
(1) $\sin A = \dfrac{5}{8}$ (2) $\sin A = \dfrac{8}{5}$ (3) $\cos A = \dfrac{5}{8}$ (4) $\cos A = \dfrac{8}{5}$ 12 _____

13. How many distinct triangles can be formed if m$\angle A = 35$, $a = 10$, and $b = 13$?
(1) 1   (2) 2   (3) 3   (4) 0   13 _____

14. When $\dfrac{3}{2}x^2 - \dfrac{1}{4}x - 4$ is subtracted from $\dfrac{5}{2}x^2 - \dfrac{3}{4}x + 1$, the difference is
(1) $-x^2 + \dfrac{1}{2}x - 5$ (2) $x^2 - \dfrac{1}{2}x + 5$ (3) $-x^2 - x - 3$ (4) $x^2 - x - 3$ 14 _____

15. The solution set of the inequality $x^2 - 3x > 10$ is
(1) $\{x \mid -2 < x < 5\}$
(2) $\{x \mid 0 < x < 3\}$
(3) $\{x \mid x < -2 \text{ or } x > 5\}$
(4) $\{x \mid x < -5 \text{ or } x > 2\}$ 15 _____

16. If $x^2 + 2 = 6x$ is solved by completing the square, an intermediate step would be
(1) $(x + 3)^2 = 7$ (2) $(x - 3)^2 = 7$ (3) $(x - 3)^2 = 11$ (4) $(x - 6)^2 = 34$ 16 _____

ALGEBRA 2 / TRIGONOMETRY
January 2011

17. Three marbles are to be drawn at random, without replacement, from a bag containing 15 red marbles, 10 blue marbles, and 5 white marbles. Which expression can be used to calculate the probability of drawing 2 red marbles and 1 white marble from the bag?

(1) $\dfrac{{}_{15}C_2 \cdot {}_5C_1}{{}_{30}C_3}$ (2) $\dfrac{{}_{15}P_2 \cdot {}_5P_1}{{}_{30}C_3}$ (3) $\dfrac{{}_{15}C_2 \cdot {}_5C_1}{{}_{30}P_3}$ (4) $\dfrac{{}_{15}P_2 \cdot {}_5P_1}{{}_{30}P_3}$ 17 ___

18. The expression $x^{-\frac{2}{5}}$ is equivalent to

(1) $-\sqrt[2]{x^5}$ (2) $-\sqrt[5]{x^2}$ (3) $\dfrac{1}{\sqrt[2]{x^5}}$ (4) $\dfrac{1}{\sqrt[5]{x^2}}$ 18 ___

19. On January 1, a share of a certain stock cost $180. Each month thereafter, the cost of a share of this stock decreased by one-third. If x represents the time, in months, and y represents the cost of the stock, in dollars, which graph best represents the cost of a share over the following 5 months?

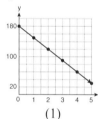
(1)

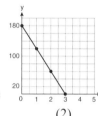
(2)

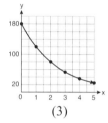
(3)

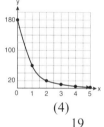
(4)

19 ___

20. In the accompanying diagram of right triangle JTM, $JT = 12$, $JM = 6$, and m$\angle JMT = 90$. What is the value of cot J?

(1) $\dfrac{\sqrt{3}}{3}$ (2) 2 (3) $\sqrt{3}$ (4) $\dfrac{2\sqrt{3}}{3}$

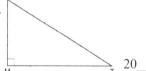

20 ___

21. For which equation does the sum of the roots equal -3 and the product of the roots equal 2?

(1) $x^2 + 2x - 3 = 0$ (3) $2x^2 + 6x + 4 = 0$
(2) $x^2 - 3x + 2 = 0$ (4) $2x^2 - 6x + 4 = 0$

21 ___

22. The expression $\dfrac{2x + 4}{\sqrt{x + 2}}$ is equivalent to

(1) $\dfrac{(2x + 4)\sqrt{x - 2}}{x - 2}$ (3) $2\sqrt{x - 2}$

(2) $\dfrac{(2x + 4)2\sqrt{x - 2}}{x - 4}$ (4) $2\sqrt{x + 2}$

22 ___

23. Which equation is sketched in the accompanying diagram?

(1) $y = \csc x$ (3) $y = \cot x$
(2) $y = \sec x$ (4) $y = \tan x$

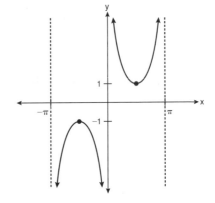

23 ____

24. The expression $\log_5\left(\dfrac{1}{25}\right)$ is equivalent to

(1) $\dfrac{1}{2}$ (2) 2 (3) $-\dfrac{1}{2}$ (4) -2

24 ____

25. A four-digit serial number is to be created from the digits 0 through 9. How many of these serial numbers can be created if 0 can *not* be the first digit, no digit may be repeated, and the last digit must be 5?
(1) 448 (2) 504 (3) 2,240 (4) 2,520

25 ____

26. Which equation represents the circle shown in the accompanying graph that passes through the point $(0, -1)$?

(1) $(x - 3)^2 + (y + 4)^2 = 16$
(2) $(x - 3)^2 + (y + 4)^2 = 18$
(3) $(x + 3)^2 + (y - 4)^2 = 16$
(4) $(x + 3)^2 + (y - 4)^2 = 18$

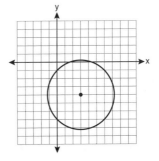

26 ____

27. Which task is *not* a component of an observational study?
(1) The researcher decides who will make up the sample.
(2) The researcher analyzes the data received from the sample.
(3) The researcher gathers data from the sample, using surveys or taking measurements.
(4) The researcher divides the sample into two groups, with one group acting as a control group.

27 ____

ALGEBRA 2 / TRIGONOMETRY
January 2011
Part II

Answer all 8 questions in this part. Each correct answer will receive 2 credits. Clearly indicate the necessary steps, including appropriate formula substitutions, diagrams, graphs, charts, etc. For all questions in this part, a correct numerical answer with no work shown will receive only 1 credit. All answers should be written in pen, except for graphs and drawings, which should be done in pencil. [16]

28. Solve algebraically for x: $16^{2x+3} = 64^{x+2}$

29. Find, to the *nearest tenth of a degree*, the angle whose measure is 2.5 radians.

30. For a given set of rectangles, the length is inversely proportional to the width. In one of these rectangles, the length is 12 and the width is 6. For this set of rectangles, calculate the width of a rectangle whose length is 9.

31. Evaluate: $10 + \sum_{n=1}^{5}(n^3 - 1)$

32. The accompanying graph represents the function $y = f(x)$.

State the domain and range of this function.

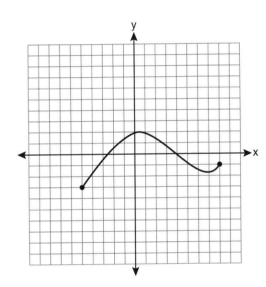

33. Express $\dfrac{\sqrt{108x^5y^8}}{\sqrt{6xy^5}}$ in simplest radical form.

34. Assume that the ages of first-year college students are normally distributed with a mean of 19 years and standard deviation of 1 year.

To the *nearest integer*, find the percentage of first-year college students who are between the ages of 18 years and 20 years, inclusive.

To the *nearest integer*, find the percentage of first-year college students who are 20 years old or older.

35. Starting with $\sin^2 A + \cos^2 A = 1$, derive the formula $\tan^2 A + 1 = \sec^2 A$.

ALGEBRA 2 / TRIGONOMETRY
January 2011
Part III

Answer all 3 questions in this part. Each correct answer will receive 4 credits. Clearly indicate the necessary steps, including appropriate formula substitutions, diagrams, graphs, charts, etc. For all questions in this part, a correct numerical answer with no work shown will receive only 1 credit. All answers should be written in pen, except for graphs and drawings, which should be done in pencil. [12]

36. Write the binomial expansion of $(2x - 1)^5$ as a polynomial in simplest form.

37. In $\triangle ABC$, $m\angle A = 32$, $a = 12$, and $b = 10$. Find the measures of the missing angles and side of $\triangle ABC$. Round each measure to the *nearest tenth*.

38. The probability that the Stormville Sluggers will win a baseball game is $\frac{2}{3}$. Determine the probability, to the *nearest thousandth*, that the Stormville Sluggers will win *at least* 6 of their next 8 games.

ALGEBRA 2 / TRIGONOMETRY
January 2011

Part IV

Answer the question in this part. A correct answer will receive 6 credits. Clearly indicate the necessary steps, including appropriate formula substitutions, diagrams, graphs, charts, etc. A correct numerical answer with no work shown will receive only 1 credit. The answer should be written in pen. [6]

39. The temperature, T, of a given cup of hot chocolate after it has been cooling for t minutes can best be modeled by the function below, where T_0 is the temperature of the room and k is a constant.

$$\ln(T - T_0) = -kt + 4.718$$

A cup of hot chocolate is placed in a room that has a temperature of 68°. After 3 minutes, the temperature of the hot chocolate is 150°. Compute the value of k to the *nearest thousandth*. [Only an algebraic solution can receive full credit.]

Using this value of k, find the temperature, T. of this cup of hot chocolate if it has been sitting in this room for a total of 10 minutes. Express your answer to the *nearest degree*. [Only an algebraic solution can receive full credit.]

ALGEBRA 2 / TRIGONOMETRY
June 2011
Part I

Answer all 27 questions in this part. Each correct answer will receive 2 credits. No partial credit will be allowed. For each question, write in the space provided the numeral preceding the word or expression that best completes the statement or answers the question. [54]

1. A doctor wants to test the effectiveness of a new drug on her patients. She separates her sample of patients into two groups and administers the drug to only one of these groups. She then compares the results. Which type of study *best* describes this situation?
(1) census (2) survey (3) observation (4) controlled experiment 1 _____

2. If $f(x) = \dfrac{x}{x^2 - 16}$, what is the value of $f(-10)$?
(1) $-\dfrac{5}{2}$ (2) $-\dfrac{5}{42}$ (3) $\dfrac{5}{58}$ (4) $\dfrac{5}{18}$ 2 _____

3. An auditorium has 21 rows of seats. The first row has 18 seats, and each succeeding row has two more seats than the previous row. How many seats are in the auditorium?
(1) 540 (2) 567 (3) 760 (4) 798 3 _____

4. Expressed as a function of a positive acute angle, $\cos(-305°)$ is equal to
(1) $-\cos 55°$ (2) $\cos 55°$ (3) $-\sin 55°$ (4) $\sin 55°$ 4 _____

5. The value of x in the equation $4^{2x+5} = 8^{3x}$ is
(1) 1 (2) 2 (3) 5 (4) -10 5 _____

6. What is the value of x in the equation $\log_5 x = 4$?
(1) 1.16 (2) 20 (3) 625 (4) 1,024 6 _____

7. The expression $\sqrt[4]{16x^2 y^7}$ is equivalent to
(1) $2x^{\frac{1}{2}} y^{\frac{7}{4}}$ (2) $2x^8 y^{28}$ (3) $4x^{\frac{1}{2}} y^{\frac{7}{4}}$ (4) $4x^8 y^{28}$ 7 _____

8. Which equation is represented by the accompanying graph?
(1) $y = 5^x$
(2) $y = 0.5^x$
(3) $y = 5^{-x}$
(4) $y = 0.5^{-x}$

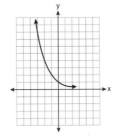

8 _____

ALGEBRA 2 / TRIGONOMETRY
June 2011

9. What is the fifteenth term of the geometric sequence $-\sqrt{5}, \sqrt{10}, -2\sqrt{5},...$?
(1) $-128\sqrt{5}$ (2) $128\sqrt{10}$ (3) $-16384\sqrt{5}$ (4) $16384\sqrt{10}$ 9 _____

10. In $\triangle ABC$, $a = 15$, $b = 14$, and $c = 13$, as shown in the accompanying diagram. What is the m$\angle C$, to the nearest degree?
(1) 53   (3) 67
(2) 59   (4) 127   10 _____

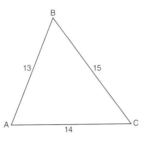

11. What is the period of the function $f(\theta) = -2\cos 3\theta$?
(1) π (2) $\dfrac{2\pi}{3}$ (3) $\dfrac{3\pi}{2}$ (4) 2π 11 _____

12. What is the range of $f(x) = (x + 4)^2 + 7$?
(1) $y \geq -4$ (2) $y \geq 4$ (3) $y = 7$ (4) $y \geq 7$ 12 _____

13. Ms. Bell's mathematics class consists of 4 sophomores, 10 juniors, and 5 seniors. How many different ways can Ms. Bell create a four-member committee of juniors if each junior has an equal chance of being selected?
(1) 210   (2) 3,876   (3) 5,040   (4) 93,024   13 _____

14. Which graph represents a relation that is *not* a function?

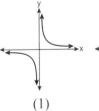 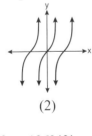 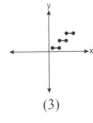 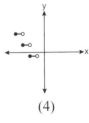

(1)   (2)   (3)   (4)   14 _____

15. The value of tan 126°43' to the *nearest ten-thousandth* is
(1) -1.3407 (2) -1.3408 (3) -1.3548 (4) -1.3549 15 _____

16. The expression $\dfrac{4}{5 - \sqrt{13}}$ is equivalent to

(1) $\dfrac{4\sqrt{13}}{5\sqrt{13} - 13}$ (2) $\dfrac{4(5 - \sqrt{13})}{38}$ (3) $\dfrac{5 + \sqrt{13}}{3}$ (4) $\dfrac{4(5 + \sqrt{13})}{38}$ 16 _____

17. Akeem invests $25,000 in an account that pays 4.75% annual interest compounded continuously. Using the formula $A = Pe^{rt}$, where A = the amount in the account after t years, P = principal invested, and r = the annual interest rate, how many years, to the *nearest tenth*, will it take for Akeem's investment to triple?
(1) 10.0 (2) 14.6 (3) 23.1 (4) 24.0 17 ___

18. The value of the expression $\sum_{r=3}^{5}(-r^2 + r)$ is
(1) −38 (2) −12 (3) 26 (4) 62 18 ___

19. Which graph shows $y = \cos^{-1} x$?

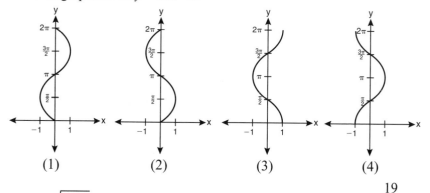

(1) (2) (3) (4)

19 ___

20. If $r = \sqrt[3]{\dfrac{A^2 B}{C}}$, then log r can be represented by

(1) $\dfrac{1}{6} \log A + \dfrac{1}{3} \log B - \log C$ (3) $\dfrac{1}{3} \log(A^2 + B) - C$

(2) $3(\log A^2 + \log B - \log C)$ (4) $\dfrac{2}{3} \log A + \dfrac{1}{3} \log B - \dfrac{1}{3} \log C$ 20 ___

21. The solution set of $\sqrt{3x+16} = x + 2$ is
(1) {−3, 4} (2) {−4, 3} (3) {3} (4) {−4} 21 ___

22. Brian correctly used a method of completing the square to solve the equation $x^2 + 7x − 11 = 0$. Brian's first step was to rewrite the equation as $x^2 + 7x = 11$. He then added a number to both sides of the equation. Which number did he add?
(1) $\dfrac{7}{2}$ (2) $\dfrac{49}{4}$ (3) $\dfrac{49}{2}$ (4) 49 22 ___

ALGEBRA 2 / TRIGONOMETRY
June 2011

23. The expression $\dfrac{\sin^2 \theta + \cos^2 \theta}{1 - \sin^2 \theta}$ is equivalent to

(1) $\cos^2 \theta$ (2) $\sin^2 \theta$ (3) $\sec^2 \theta$ (4) $\csc^2 \theta$ 23 ____

24. The number of minutes students took to complete a quiz is summarized in the table below.

Minutes	14	15	16	17	18	19	20
Number of Students	5	3	x	5	2	10	1

If the mean number of minutes was 17, which equation could be used to calculate the value of x?

(1) $17 = \dfrac{119 + x}{x}$ (3) $17 = \dfrac{446 + x}{26 + x}$

(2) $17 = \dfrac{119 + 16x}{x}$ (4) $17 = \dfrac{446 + 16x}{26 + x}$ 24 ____

25. What is the radian measure of the smaller angle formed by the hands of a clock at 7 o'clock?

(1) $\dfrac{\pi}{2}$ (2) $\dfrac{2\pi}{3}$ (3) $\dfrac{5\pi}{6}$ (4) $\dfrac{7\pi}{6}$ 25 ____

26. What is the coefficient of the fourth term in the expansion of $(a - 4b)^9$?

(1) −5,376   (2) −336   (3) 336   (4) 5,376   26 ____

27. Samantha constructs the accompanying scatter plot from a set of data. Based on her scatter plot, which regression model would be most appropriate?
(1) exponential
(2) linear
(3) logarithmic
(4) power

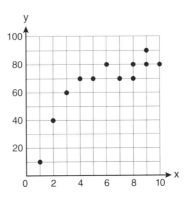

27 ____

ALGEBRA 2 / TRIGONOMETRY
June 2011
Part II

Answer all 8 questions in this part. Each correct answer will receive 2 credits. Clearly indicate the necessary steps, including appropriate formula substitutions, diagrams, graphs, charts, etc. For all questions in this part, a correct numerical answer with no work shown will receive only 1 credit. All answers should be written in pen, except for graphs and drawings, which should be done in pencil. [16]

28. Express the product of $\left(\dfrac{1}{2}y^2 - \dfrac{1}{3}y\right)$ and $\left(12y + \dfrac{3}{5}\right)$ as a trinomial.

29. In a study of 82 video game players, the researchers found that the ages of these players were normally distributed, with a mean age of 17 years and a standard deviation of 3 years. Determine if there were 15 video game players in this study over the age of 20. Justify your answer.

30. Write a quadratic equation such that the sum of its roots is 6 and the product of its roots is −27.

ALGEBRA 2 / TRIGONOMETRY
June 2011

1. Evaluate $e^{x \ln y}$ when $x = 3$ and $y = 2$.

2. If $f(x) = x^2 - 6$, find $f^{-1}(x)$.

33. Factor the expression $12t^8 - 75t^4$ completely.

34. Simplify the expression $\dfrac{3x^{-4}y^5}{(2x^3y^{-7})^{-2}}$ and write the answer using only positive exponents.

35. If $f(x) = x^2 - 6$ and $g(x) = 2^x - 1$, determine the value of $(g \circ f)(-3)$.

Part III

Answer all 3 questions in this part. Each correct answer will receive 4 credits. Clearly indicate the necessary steps, including appropriate formula substitutions, diagrams, graphs, charts, etc. For all questions in this part, a correct numerical answer with no work shown will receive only 1 credit. All answers should be written in pen, except for graphs and drawings, which should be done in pencil. [12]

36. Express as a single fraction the exact value of $\sin 75°$.

37. Solve the inequality $-3|6 - x| < -15$ for x. Graph the solution on the line below.

38. The probability that a professional baseball player will get a hit is $\frac{1}{3}$. Calculate the exact probability that he will get *at least* 3 hits in 5 attempts.

Part IV

Answer the question in this part. A correct answer will receive 6 credits. Clearly indicate the necessary steps, including appropriate formula substitutions, diagrams, graphs, charts, etc. A correct numerical answer with no work shown will receive only 1 credit. The answer should be written in pen. [6]

39. Solve the following system of equations algebraically:
$$5 = y - x$$
$$4x^2 = -17x + y + 4$$

ALGEBRA 2 / TRIGONOMETRY
January 2012
Part I

Answer all 27 questions in this part. Each correct answer will receive 2 credits. No partial credit will be allowed. For each question, write in the space provided the numeral preceding the word or expression that best completes the statement or answers the question. [54]

1. The yearbook staff has designed a survey to learn student opinions on how the yearbook could be improved for this year. If they want to distribute this survey to 100 students and obtain the most reliable data, they should survey
(1) every third student sent to the office
(2) every third student to enter the library
(3) every third student to enter the gym for the basketball game
(4) every third student arriving at school in the morning

2. What is the sum of the first 19 terms of the sequence 3, 10, 17, 24, 31, ...?
(1) 1188 (2) 1197 (3) 1254 (4) 1292

3. Which expression, when rounded to three decimal places, is equal to -1.155?
(1) $\sec\left(\dfrac{5\pi}{6}\right)$ (2) $\tan(49°20')$ (3) $\sin\left(-\dfrac{3\pi}{5}\right)$ (4) $\csc(-118°)$

4. If $f(x) = 4x - x^2$ and $g(x) = \dfrac{1}{x}$, then $(f \circ g)\left(\dfrac{1}{2}\right)$ is equal to
(1) $\dfrac{4}{7}$ (2) -2 (3) $\dfrac{7}{2}$ (4) 4

5. A population of rabbits doubles every 60 days according to the formula $P = 10(2)^{\frac{t}{60}}$, where P is the population of rabbits on day t. What is the value of t when the population is 320?
(1) 240 (2) 300 (3) 660 (4) 960

6. What is the product of $\left(\dfrac{x}{4} - \dfrac{1}{3}\right)$ and $\left(\dfrac{x}{4} + \dfrac{1}{3}\right)$?
(1) $\dfrac{x^2}{8} - \dfrac{1}{9}$ (2) $\dfrac{x^2}{16} - \dfrac{1}{9}$ (3) $\dfrac{x^2}{8} - \dfrac{x}{6} - \dfrac{1}{9}$ (4) $\dfrac{x^2}{16} - \dfrac{x}{6} - \dfrac{1}{9}$

7. Which is a graph of $y = \cot x$?

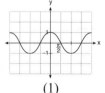 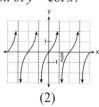 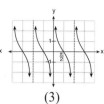 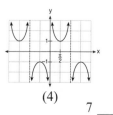

(1)　　　　　(2)　　　　　(3)　　　　　(4)

ALGEBRA 2 / TRIGONOMETRY
January 2012

8. Which expression always equals 1?
(1) $\cos^2 x - \sin^2 x$
(2) $\cos^2 x + \sin^2 x$
(3) $\cos x - \sin x$
(4) $\cos x + \sin x$

8 _____

9. What are the sum and product of the roots of the equation $6x^2 - 4x - 12 = 0$?
(1) sum $= -\frac{2}{3}$; product $= -2$
(2) sum $= \frac{2}{3}$; product $= -2$
(3) sum $= -2$; product $= \frac{2}{3}$
(4) sum $= -2$; product $= -\frac{2}{3}$

9 _____

10. Given $\triangle ABC$ with $a = 9$, $b = 10$, and $m\angle B = 70$, what type of triangle can be drawn?
(1) an acute triangle, only
(2) an obtuse triangle, only
(3) both an acute triangle and an obtuse triangle
(4) neither an acute triangle nor an obtuse triangle

10 _____

11. When $x^{-1} + 1$ is divided by $x + 1$, the quotient equals
(1) 1
(2) $\frac{1}{x}$
(3) x
(4) $-\frac{1}{x}$

11 _____

12. If the amount of time students work in any given week is normally distributed with a mean of 10 hours per week and a standard deviation of 2 hours, what is the probability a student works between 8 and 11 hours per week?
(1) 34.1% (2) 38.2% (3) 53.2% (4) 68.2%

12 _____

13. What is the conjugate of $\frac{1}{2} + \frac{3}{2}i$?
(1) $-\frac{1}{2} + \frac{3}{2}i$
(2) $\frac{1}{2} - \frac{3}{2}i$
(3) $\frac{3}{2} + \frac{1}{2}i$
(4) $-\frac{1}{2} - \frac{3}{2}i$

13 _____

14. Given angle A in Quadrant I with $\sin A = \frac{12}{13}$ and angle B in Quadrant II with $\cos B = -\frac{3}{5}$, what is the value of $\cos(A - B)$?
(1) $\frac{33}{65}$
(2) $-\frac{33}{65}$
(3) $\frac{63}{65}$
(4) $-\frac{63}{65}$

14 _____

15. Which expression represents the third term in the expansion of $(2x^4 - y)^3$?
(1) $-y^3$ (2) $-6x^4y^2$ (3) $6x^4y^2$ (4) $2x^4y^2$

15 _____

16. What is the solution set of the equation $3x^5 - 48x = 0$?
(1) $\{0, \pm 2\}$ (2) $\{0, \pm 2, 3\}$ (3) $\{0, \pm 2, \pm 2i\}$ (4) $\{\pm 2, \pm 2i\}$

16 _____

17. A sequence has the following terms: $a_1 = 4$, $a_2 = 10$, $a_3 = 25$, $a_4 = 62.5$. Which formula represents the nth term in this sequence?
(1) $a_n = 4 + 2.5n$
(2) $a_n = 4 + 2.5(n - 1)$
(3) $a_n = 4(2.5)^n$
(4) $a_n = 4(2.5)^{n-1}$

17 _____

18. In parallelogram $BFLO$, $OL = 3.8$, $LF = 7.4$, and $m\angle O = 126$. If diagonal $\overline{BL}$ is drawn, what is the area of $\triangle BLF$?
(1) 11.4 (2) 14.1 (3) 22.7 (4) 28.1   18 _____

19. Which statement about the graph of the equation $y = e^x$ is *not* true?
(1) It is asymptotic to the x-axis.
(2) The domain is the set of all real numbers.
(3) It lies in Quadrants I and II.
(4) It passes through the point $(e, 1)$.   19 _____

20. What is the number of degrees in an angle whose measure is 2 radians?
(1) $\dfrac{360}{\pi}$ (2) $\dfrac{\pi}{360}$ (3) 360 (4) 90   20 _____

21. A spinner is divided into eight equal sections. Five sections are red and three are green. If the spinner is spun three times, what is the probability that it lands on red *exactly* twice?
(1) $\dfrac{25}{64}$ (2) $\dfrac{45}{512}$ (3) $\dfrac{75}{512}$ (4) $\dfrac{225}{512}$ 21 _____

22. What is the range of $f(x) = |x - 3| + 2$?
(1) $\{x | x \geq 3\}$ (3) $\{x | x \in \text{real numbers}\}$
(2) $\{y | y \geq 2\}$ (4) $\{y | y \in \text{real numbers}\}$ 22 _____

23. Which calculator output shows the strongest linear relationship between x and y?

(1) Lin Reg   (2) Lin Reg   (3) Lin Reg   (4) Lin Reg
$y = a + bx$ $y = a + bx$ $y = a + bx$ $y = a + bx$
$a = 59.026$ $a = .7$ $a = 2.45$ $a = -2.9$
$b = 6.767$ $b = 24.2$ $b = .95$ $b = 24.1$
$r = .8643$ $r = .8361$ $r = .6022$ $r = -.8924$ 23 _____

24. If $\log x^2 - \log 2a = \log 3a$, then $\log x$ expressed in terms of $\log a$ is equivalent to
(1) $\dfrac{1}{2} \log 5a$ (3) $\log 6 + \log a$
(2) $\dfrac{1}{2} \log 6 + \log a$ (4) $\log 6 + 2 \log a$ 24 _____

25. Which function is one-to-one?
(1) $f(x) = |x|$ (2) $f(x) = 2^x$ (3) $f(x) = x^2$ (4) $f(x) = \sin x$ 25 _____

26. If p varies inversely as q, and $p = 10$ when $q = \dfrac{3}{2}$, what is the value of p when $q = \dfrac{3}{5}$?
(1) 25 (2) 15 (3) 9 (4) 4   26 _____

ALGEBRA 2 / TRIGONOMETRY
January 2012

27. Which equation is graphed in the diagram below?

(1) $y = 3\cos\left(\dfrac{\pi}{30}x\right) + 8$

(2) $y = 3\cos\left(\dfrac{\pi}{15}x\right) + 5$

(3) $y = -3\cos\left(\dfrac{\pi}{30}x\right) + 8$

(4) $y = -3\cos\left(\dfrac{\pi}{15}x\right) + 5$

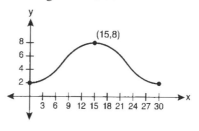

27 _____

Part II

Answer all 8 questions in this part. Each correct answer will receive 2 credits. Clearly indicate the necessary steps, including appropriate formula substitutions, diagrams, graphs, charts, etc. For all questions in this part, a correct numerical answer with no work shown will receive only 1 credit. All answers should be written in pen, except for graphs and drawings, which should be done in pencil. [16]

28. Find the solution of the inequality $x^2 - 4x > 5$, algebraically.

29. Solve algebraically for x: $4 - \sqrt{2x - 5} = 1$

30. Evaluate: $\displaystyle\sum_{n=1}^{3}(-n^4 - n)$

31. Express in simplest form: $\sqrt[3]{\dfrac{a^6 b^9}{-64}}$

32. A blood bank needs twenty people to help with a blood drive. Twenty-five people have volunteered. Find how many different groups of twenty can be formed from the twenty-five volunteers.

33. On the axes below, for $-2 \le x \le 2$, graph $y = 2^{x+1} - 3$.

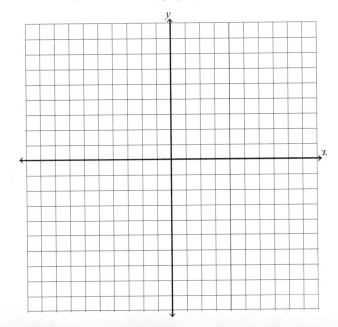

34. Write an equation of the circle shown in the diagram below.

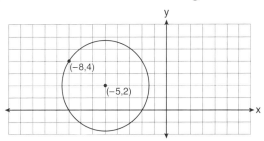

35. Express the exact value of csc 60°, with a rational denominator.

Part III

Answer all 3 questions in this part. Each correct answer will receive 4 credits. Clearly indicate the necessary steps, including appropriate formula substitutions, diagrams, graphs, charts, etc. For all questions in this part, a correct numerical answer with no work shown will receive only 1 credit. All answers should be written in pen, except for graphs and drawings, which should be done in pencil. [12]

36. The diagram below shows the plans for a cell phone tower. A guy wire attached to the top of the tower makes an angle of 65 degrees with the ground. From a point on the ground 100 feet from the end of the guy wire, the angle of elevation to the top of the tower is 32 degrees. Find the height of the tower, to the *nearest foot*.

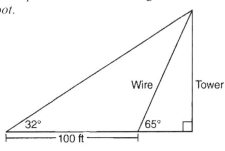

37. If $\log_4 x = 2.5$ and $\log_y 125 = -\dfrac{3}{2}$, find the numerical value of $\dfrac{x}{y}$, in simplest form.

38. A population of single-celled organisms was grown in a Petri dish over a period of 16 hours. The number of organisms at a given time is recorded in the accompanying table.

Determine the exponential regression equation model for these data, rounding all values to the *nearest ten-thousandth*.

Using this equation, predict the number of single-celled organisms, to the *nearest whole number*, at the end of the 18th hour.

Time, hrs (x)	Number of Organisms (y)
0	25
2	36
4	52
6	68
8	85
10	104
12	142
16	260

ALGEBRA 2 / TRIGONOMETRY
January 2012
Part IV

Answer the question in this part. A correct answer will receive 6 credits. Clearly indicate the necessary steps, including appropriate formula substitutions, diagrams, graphs, charts, etc. A correct numerical answer with no work shown will receive only 1 credit. The answer should be written in pen. [6]

39. Perform the indicated operations and simplify completely:

$$\frac{x^3 - 3x^2 + 6x - 18}{x^2 - 4x} \cdot \frac{2x - 4}{x^4 - 3x^3} \div \frac{x^2 + 2x - 8}{16 - x^2}$$

ALGEBRA 2 / TRIGONOMETRY
June 2012
Part I

Answer all 27 questions in this part. Each correct answer will receive 2 credits. No partial credit will be allowed. For each question, write in the space provided the numeral preceding the word or expression that best completes the statement or answers the question. [54]

1. What is the product of $\left(\frac{2}{5}x - \frac{3}{4}y^2\right)$ and $\left(\frac{2}{5}x + \frac{3}{4}y^2\right)$?

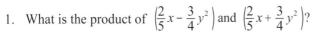

(1) $\frac{4}{25}x^2 - \frac{9}{16}y^4$ (2) $\frac{4}{25}x - \frac{9}{16}y^2$ (3) $\frac{2}{5}x^2 - \frac{3}{4}y^4$ (4) $\frac{4}{5}x$

1 _____

2. What is the domain of the function shown?
(1) $-1 \leq x \leq 6$
(2) $-1 \leq y \leq 6$
(3) $-2 \leq x \leq 5$
(4) $-2 \leq y \leq 5$

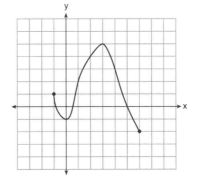

2 _____

3. What is the solution set for $2\cos\theta - 1 = 0$ in the interval $0° \leq \theta < 360°$?
(1) {30°, 150°}   (3) {30°, 330°}
(2) {60°, 120°}   (4) {60°, 300°}

3 _____

4. The expression $\sqrt[3]{64a^{16}}$ is equivalent to
(1) $8a^4$ (2) $8a^8$ (3) $4a^5\sqrt[3]{a}$ (4) $4a\sqrt[3]{a^5}$

4 _____

5. Which summation represents $5 + 7 + 9 + 11 + \ldots + 43$?

(1) $\sum_{n=5}^{43} n$ (2) $\sum_{n=1}^{20}(2n+3)$ (3) $\sum_{n=4}^{24}(2n-3)$ (4) $\sum_{n=3}^{23}(3n-4)$

5 _____

6. If $m\angle\theta = -50$, which diagram represents θ drawn in standard position?

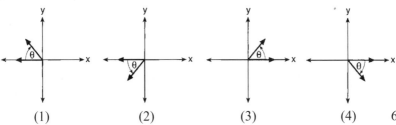

(1)   (2)   (3)   (4)

6 _____

ALGEBRA 2 / TRIGONOMETRY
June 2012

7. If $\log_b x = 3\log_b p - (2\log_b t + \frac{1}{2}\log_b r)$, then the value of x is

(1) $\dfrac{p^3}{\sqrt{t^2 r}}$ (2) $p^3 t^2 r^{\frac{1}{2}}$ (3) $\dfrac{p^3 t^2}{\sqrt{r}}$ (4) $\dfrac{p^3}{t^2 \sqrt{r}}$ 7 ____

8. Which equation has roots with the sum equal to $\dfrac{9}{4}$ and the product equal to $\dfrac{3}{4}$?

(1) $4x^2 + 9x + 3 = 0$ (3) $4x^2 - 9x + 3 = 0$
(2) $4x^2 + 9x - 3 = 0$ (4) $4x^2 - 9x - 3 = 0$ 8 ____

9. Which graph represents the solution set of $\left|\dfrac{4x-5}{3}\right| > 1$?

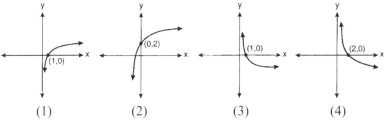

9 ____

10. Which expression is equivalent to $\dfrac{x^{-1} y^4}{3x^{-5} y^{-1}}$?

(1) $\dfrac{x^4 y^5}{3}$ (2) $\dfrac{x^5 y^4}{3}$ (3) $3x^4 y^5$ (4) $\dfrac{y^4}{3x^5}$ 10 ____

11. Which graph represents the function $\log_2 x = y$?

(1)   (2)   (3)   (4)   11 ____

12. A circle is drawn to represent a pizza with a 12 inch diameter. The circle is cut into eight congruent pieces. What is the length of the outer edge of anyone piece of this circle?

(1) $\dfrac{3\pi}{4}$ (2) π (3) $\dfrac{3\pi}{2}$ (4) 3π 12 ____

13. What is the solution set for the equation $\sqrt{5x+29} = x+3$?

(1) $\{4\}$ (2) $\{-5\}$ (3) $\{4, 5\}$ (4) $\{-5, 4\}$ 13 ____

14. When factored completely, $x^3 + 3x^2 - 4x - 12$ equals

(1) $(x+2)(x-2)(x-3)$ (3) $(x^2-4)(x+3)$
(2) $(x+2)(x-2)(x+3)$ (4) $(x^2-4)(x-3)$ 14 ____

15. What is the middle term in the expansion of $\left(\dfrac{x}{2} - 2y\right)^6$?

(1) $20x^3y^3$ (2) $-\dfrac{15}{4}x^4y^2$ (3) $-20x^3y^3$ (4) $\dfrac{15}{4}x^4y^2$ 15 ____

16. Which expression is equivalent to $(n \circ m \circ p)(x)$, given $m(x) = \sin x$, $n(x) = 3x$, and $p(x) = x^2$?

(1) $\sin(3x)^2$ (2) $3\sin x^2$ (3) $\sin^2(3x)$ (4) $3\sin^2 x$ 16 ____

17. The value of $\csc 138°23'$ rounded to four decimal places is

(1) -1.3376 (2) -1.3408 (3) 1.5012 (4) 1.5057 17 ____

18. Which function is one-to-one?

(1) $k(x) = x^2 + 2$
(2) $g(x) = x^3 + 2$
(3) $f(x) = |x| + 2$
(4) $j(x) = x^4 + 2$ 18 ____

19. The conjugate of the complex expression $-5x + 4i$ is

(1) $5x - 4i$ (2) $5x + 4i$ (3) $-5x - 4i$ (4) $-5x + 4i$ 19 ____

20. What is a positive value of $\tan \dfrac{1}{2}x$, when $\sin x = 0.8$?

(1) 0.5 (2) 0.4 (3) 0.33 (4) 0.25 20 ____

21. The table below displays the results of a survey regarding the number of pets each student in a class has. The average number of pets per student in this class is 2.

Number of Pets	0	1	2	3	4	5
Number of Students	4	6	10	0	k	2

What is the value of k for this table?

(1) 9 (2) 2 (3) 8 (4) 4   21 ____

22. How many negative solutions to the equation $2x^3 - 4x^2 + 3x - 1 = 0$ exist?

(1) 1 (2) 2 (3) 3 (4) 0   22 ____

23. A study finds that 80% of the local high school students text while doing homework. Ten students are selected at random from the local high school. Which expression would be part of the process used to determine the probability that, *at most*, 7 of the 10 students text while doing homework?

(1) $_{10}C_6 \left(\dfrac{4}{5}\right)^6 \left(\dfrac{1}{5}\right)^4$

(2) $_{10}C_7 \left(\dfrac{4}{5}\right)^{10} \left(\dfrac{1}{5}\right)^7$

(3) $_{10}C_8 \left(\dfrac{7}{10}\right)^{10} \left(\dfrac{3}{10}\right)^2$

(4) $_{10}C_9 \left(\dfrac{7}{10}\right)^9 \left(\dfrac{3}{10}\right)^1$ 23 ____

ALGEBRA 2 / TRIGONOMETRY
June 2012

24. In which interval of f(x) = cos(x) is the inverse also a function?
(1) $-\frac{\pi}{2} < x < \frac{\pi}{2}$
(3) $0 \le x \le \pi$
(2) $-\frac{\pi}{2} \le x \le \frac{\pi}{2}$
(4) $\frac{\pi}{2} \le x \le \frac{3\pi}{2}$

24 _____

25. As shown in the table, a person's target heart rate during exercise changes as the person gets older. Which value represents the linear correlation coefficient, rounded to the *nearest thousandth*, between a person's age, in years, and that person's target heart rate, in beats per minute?

Age (years)	Target Heart Rate (beats per minute)
20	135
25	132
30	129
35	125
40	122
45	119
50	115

(1) –0.999 (3) 0.998
(2) –0.664 (4) 1.503

25 _____

26. In $\triangle MNP$, m = 6 and n = 10. Two distinct triangles can be constructed if the measure of angle M is
(1) 35 (2) 40 (3) 45 (4) 50

26 _____

27. If order does *not* matter, which selection of students would produce the most possible committees?
(1) 5 out of 15 (2) 5 out of 25 (3) 20 out of 25 (4) 15 out of 25

27 _____

Part II
Answer all 8 questions in this part. Each correct answer will receive 2 credits. Clearly indicate the necessary steps, including appropriate formula substitutions, diagrams, graphs, charts, etc. For all questions in this part, a correct numerical answer with no work shown will receive only 1 credit. All answers should be written in pen, except for graphs and drawings, which should be done in pencil. [16]

28. Determine the value of *n* in simplest form: $i^{13} + i^{18} + i^{31} + n = 0$

29. The formula for continuously compounded interest is $A = Pe^{rt}$, where A is the amount of money in the account, P is the initial investment, r is the interest rate, and t is the time in years.

Using the formula, determine, to the *nearest dollar*, the amount in the account after 8 years if $750 is invested at an annual rate of 3%.

30. Express $\cos\theta\,(\sec\theta - \cos\theta)$, in terms of $\sin\theta$.

31. A cup of soup is left on a countertop to cool. The table below gives the temperatures, in degrees Fahrenheit, of the soup recorded over a 10-minute period.

Time in Minutes (x)	0	2	4	6	8	10
Temperature in °F (y)	180.2	165.8	146.3	135.4	127.7	110.5

Write an exponential regression equation for the data, rounding all values to the *nearest thousandth*.

32. Find, to the *nearest tenth*, the radian measure of 216°.

33. Find the third term in the recursive sequence $a_{k+1} = 2a_k - 1$, where $a_1 = 3$.

34. The two sides and included angle of a parallelogram are 18, 22, and 60°. Find its exact area in simplest form.

35. Write an equation for the graph of the trigonometric function shown below.

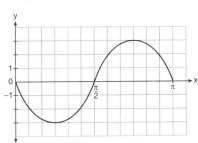

ALGEBRA 2 / TRIGONOMETRY
June 2012

Part III

Answer all 3 questions in this part. Each correct answer will receive 4 credits. Clearly indicate the necessary steps, including appropriate formula substitutions, diagrams, graphs, charts, etc. For all questions in this part, a correct numerical answer with no work shown will receive only 1 credit. All answers should be written in pen, except for graphs and drawings, which should be done in pencil. [12]

36. Express in simplest form: $\dfrac{\dfrac{4-x^2}{x^2+7x+12}}{\dfrac{2x-4}{x+3}}$

37. During a particular month, a local company surveyed all its employees to determine their travel times to work, in minutes. The data for all 15 employees are shown below.

$$\begin{array}{ccccc} 25 & 55 & 40 & 65 & 29 \\ 45 & 59 & 35 & 25 & 37 \\ 52 & 30 & 8 & 40 & 55 \end{array}$$

Determine the number of employees whose travel time is within one standard deviation of the mean.

38. The measures of the angles between the resultant and two applied forces are 60° and 45°, and the magnitude of the resultant is 27 pounds. Find, to the *nearest pound*, the magnitude of each applied force.

Part IV
Answer the question in this part. A correct answer will receive 6 credits. Clearly indicate the necessary steps, including appropriate formula substitutions, diagrams, graphs, charts, etc. A correct numerical answer with no work shown will receive only 1 credit. The answer should be written in pen. [6]

39. Solve algebraically for all values of x: $\quad 81^{x^3+2x^2} = 27^{\frac{5x}{3}}$

ALGEBRA 2 / TRIGONOMETRY
January 2013
Part I

Answer all 27 questions in this part. Each correct answer will receive 2 credits. No partial credit will be allowed. For each question, write in the space provided the numeral preceding the word or expression that best completes the statement or answers the question. [54]

1. What is the equation of the graph shown to the right?
   (1) $y = 2^x$
   (2) $y = 2^{-x}$
   (3) $x = 2^y$
   (4) $x = 2^{-y}$

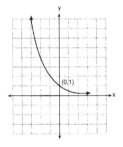

1 _____

2. Which ordered pair is a solution of the system of equations shown below?
$$x + y = 5$$
$$(x + 3)^2 + (y - 3)^2 = 53$$

(1) (2, 3)   (2) (5, 0)   (3) (–5, 10)   (4) (–4, 9)   2 _____

3. The relationship between t, a student's test scores, and d, the student's success in college, is modeled by the equation $d = 0.48t + 75.2$. Based on this linear regression model, the correlation coefficient could be
(1) between -1 and 0
(2) between 0 and 1
(3) equal to -1
(4) equal to 0

3 _____

4. What is the common ratio of the geometric sequence shown below?
$$-2, 4, -8, 16, \dots$$
(1) $-\dfrac{1}{2}$ (2) 2   (3) –2   (4) –6   4 _____

5. Given the relation $\{(8, 2), (3, 6), (7, 5), (k, 4)\}$, which value of k will result in the relation *not* being a function?
(1) 1   (2) 2   (3) 3   (4) 4   5 _____

6. Which expression is equivalent to $(9x^2 y^6)^{-\frac{1}{2}}$?
(1) $\dfrac{1}{3xy^3}$ (2) $3xy^3$ (3) $\dfrac{3}{xy^3}$ (4) $\dfrac{xy^3}{3}$ 6 _____

7. In a certain high school, a survey revealed the mean amount of bottled water consumed by students each day was 153 bottles with a standard deviation of 22 bottles. Assuming the survey represented a normal distribution, what is the range of the number of bottled waters that approximately 68.2% of the students drink?
(1) 131-164   (2) 131-175   (3) 142-164   (4) 142-175   7 _____

ALGEBRA 2 / TRIGONOMETRY
January 2013

8. What is the fourth term in the binomial expansion $(x-2)^8$?
(1) $448x^5$　　(2) $448x^4$　　(3) $-448x^5$　　(4) $-448x^4$　　8 _____

9. Which value of k satisfies the equation $8^{3k+4} = 4^{2k-1}$?
(1) -1　　(2) $-\dfrac{9}{4}$　　(3) -2　　(4) $-\dfrac{14}{5}$　　9 _____

10. There are eight people in a tennis club. Which expression can be used to find the number of different ways they can place first, second, and third in a tournament?
(1) $_8P_3$　　(2) $_8C_3$　　(3) $_8P_5$　　(4) $_8C_5$　　10 _____

11. If $\sin A = \dfrac{1}{3}$, what is the value of $\cos 2A$?
(1) $-\dfrac{2}{3}$　　(2) $\dfrac{2}{3}$　　(3) $-\dfrac{7}{9}$　　(4) $\dfrac{7}{9}$　　11 _____

12. In the interval $0° \le x < 360°$, $\tan x$ is undefined when x equals
(1) 0° and 90°　　(3) 180° and 270°
(2) 90° and 180°　　(4) 90° and 270°　　12 _____

13. If $f(x) = \sqrt{9-x^2}$, what are its domain and range?
(1) domain: $\{x \mid -3 \le x \le 3\}$; range: $\{y \mid 0 \le y \le 3\}$
(2) domain: $\{x \mid x \ne \pm 3\}$; range: $\{y \mid 0 \le y \le 3\}$
(3) domain: $\{x \mid x \le -3 \text{ or } x \ge 3\}$; range: $\{y \mid y \ne 0\}$
(4) domain: $\{x \mid x \ne 3\}$; range: $\{y \mid y \ge 0\}$　　13 _____

14. When $x^2 + 3x - 4$ is subtracted from $x^3 + 3x^2 - 2x$, the difference is
(1) $x^3 + 2x^2 - 5x + 4$　　(3) $-x^3 + 4x^2 + x - 4$
(2) $x^3 + 2x^2 + x - 4$　　(4) $-x^3 - 2x^2 + 5x + 4$　　14 _____

15. In the accompanying diagram, the length of which line segment is equal to the exact value of $\sin \theta$?

(1) $\overline{TO}$　　(3) $\overline{OR}$
(2) $\overline{TS}$　　(4) $\overline{OS}$

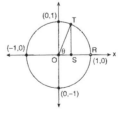

15 _____

16. The area of triangle ABC is 42. If $AB = 8$ and $m\angle B = 61$, the length of $\overline{BC}$ is approximately
(1) 5.1　　(2) 9.2　　(3) 12.0　　(4) 21.7　　16 _____

17. When factored completely, the expression $3x^3 - 5x^2 - 48x + 80$ is equivalent to
(1) $(x^2 - 16)(3x - 5)$　　(3) $(x+4)(x-4)(3x-5)$
(2) $(x^2 + 16)(3x - 5)(3x + 5)$　　(4) $(x+4)(x-4)(3x-5)(3x-5)$　　17 _____

18. The value of sin (180 + x) is equivalent to
(1) – sin x   (2) – sin (90 – x)   (3) sin x   (4) sin (90 – x)   18 ___

19. The sum of $\sqrt[3]{6a^4b^2}$ and $\sqrt[3]{162a^4b^2}$, expressed in simplest radical form, is
(1) $\sqrt[6]{168a^8b^4}$
(2) $2a^2b\sqrt[3]{21a^2b}$
(3) $4a\sqrt[3]{6ab^2}$
(4) $10a^2b\sqrt[3]{8}$ 19 ___

20. Which equation is represented by the accompanying graph?
(1) $y = 2 \cos 3x$ (3) $y = 2 \cos \frac{2\pi}{3}x$
(2) $y = 2 \sin 3x$ (4) $y = 2 \sin \frac{2\pi}{3}x$ 20 ___

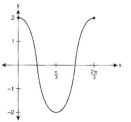

21. The quantities p and q vary inversely. If p = 20 when q = –2, and p = x when q = –2x + 2, then x equals
(1) –4 and 5   (2) $\frac{20}{19}$ (3) –5 and 4   (4) $-\frac{1}{4}$ 21 ___

22. What is the solution set of the equation $-\sqrt{2} \sec x = 2$ when $0° \leq x < 360°$?
(1) {45°, 135°, 225°, 315°}   (3) {135°, 225°}
(2) {45°, 315°}   (4) {225°, 315°}   22 ___

23. The discriminant of a quadratic equation is 24. The roots are
(1) imaginary   (3) real, rational, and unequal
(2) real, rational, and equal   (4) real, irrational, and unequal   23 ___

24. How many different six-letter arrangements can be made using the letters of the word "TATTOO"?
(1) 60   (2) 90   (3) 120   (4) 720   24 ___

25. Expressed in simplest form, $\frac{3y}{2y-6} + \frac{9}{6-2y}$ is equivalent to
(1) $\frac{-6y^2 + 36y - 54}{(2y-6)(6-2y)}$ (2) $\frac{3y-9}{2y-6}$ (3) $\frac{3}{2}$ (4) $-\frac{3}{2}$ 25 ___

26. If log 2 = a and log 3 = b, the expression $\log \frac{9}{20}$ is equivalent to
(1) $2b - a + 1$ (3) $b^2 - a + 10$
(2) $2b - a - 1$ (4) $\frac{2b}{a+1}$ 26 ___

27. The expression $(x + i)^2 - (x - i)^2$ is equivalent to
(1) 0   (2) –2   (3) $-2 + 4xi$ (4) $4xi$ 27 ___

ALGEBRA 2 / TRIGONOMETRY
January 2013
Part II

Answer all 8 questions in this part. Each correct answer will receive 2 credits. Clearly indicate the necessary steps, including appropriate formula substitutions, diagrams, graphs, charts, etc. For all questions in this part, a correct numerical answer with no work shown will receive only 1 credit. All answers should be written in the space provided.

28. Determine the sum of the first twenty terms of the sequence whose first five terms are 5, 14, 23, 32 and 41.

29. Determine the sum and the product of the roots of $3x^2 = 11x - 6$.

30. If $\sec(a + 15)° = \csc(2a)°$, find the smallest positive value of a, in degrees.

31. The heights, in inches, of 10 high school varsity basketball players are 78, 79, 79, 72, 75, 71, 74, 74, 83, and 71. Find the interquartile range of this data set.

32. Solve the equation $6x^2 - 2x - 3 = 0$ and express the answer in simplest radical form.

33. The number of bacteria present in a Petri dish can be modeled by the function $N = 50e^{3t}$, where N is the number of bacteria present in the Petri dish after t hours. Using this model, determine, to the *nearest hundredth*, the number of hours it will take for N to reach 30,700.

34. Determine the solution of the inequality $|3 - 2x| \geq 7$.
[The use of the grid below is optional.]

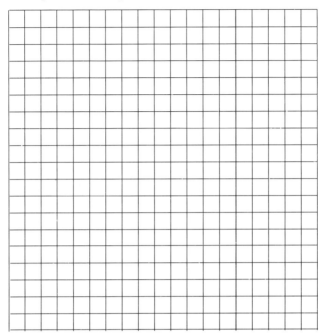

35. Convert 3 radians to degrees and express the answer to the *nearest minute*.

ALGEBRA 2 / TRIGONOMETRY
January 2013
Part III

Answer all 3 questions in this part. Each correct answer will receive 4 credits. Clearly indicate the necessary steps, including appropriate formula substitutions, diagrams, graphs, charts, etc. For all questions in this part, a correct numerical answer with no work shown will receive only 1 credit. All answers should be written in the space provided. [12]

36. Solve algebraically for all values of x:

$$\log_{(x+4)}(17x-4) = 2$$

37. The data collected by a biologist showing the growth of a colony of bacteria at the end of each hour are displayed in the table below.

Time, hour, (x)	0	1	2	3	4	5
Population (y)	250	330	580	800	1650	3000

Write an exponential regression equation to model these data. Round *all* values to the *nearest thousandth*.

Assuming this trend continues, use this equation to estimate, to the *nearest ten*, the number of bacteria in the colony at the end of 7 hours.

38. As shown in the diagram below, fire-tracking station A is 100 miles due west of fire-tracking station B. A forest fire is spotted at F, on a bearing 47° northeast of station A and 15° northeast of station B. Determine, to the *nearest tenth of a mile*, the distance the fire is from *both* station A and station B. [N represents due north.]

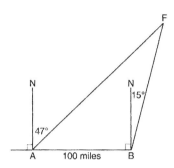

Part IV

Answer the question in this part. **A correct answer will receive 6 credits. Clearly indicate the necessary steps, including appropriate formula substitutions, diagrams, graphs, charts, etc. A correct numerical answer with no work shown will receive only 1 credit. The answer should be written in the space provided.** [6]

39. Solve algebraically for x: $\sqrt{x^2 + x - 1} + 11x = 7x + 3$